Introduction to Bluetooth,
Technology, Market, Operation, Profiles, and Services

Lawrence Harte

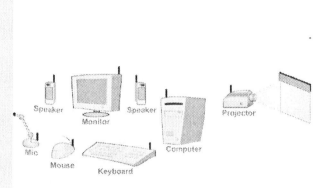

Bluetooth Devices

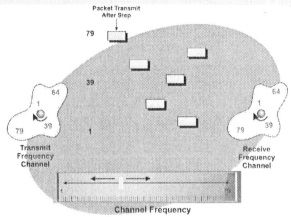

Bluetooth Operation

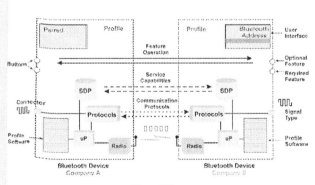

Profiles

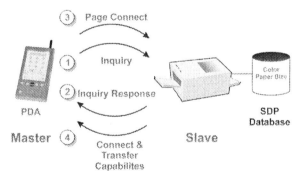

Serivce Discovery

Excerpted From:

Wireless Systems

With Updated Information

ALTHOS Publishing

ALTHOS Publishing

About the Authors

 Mr. Harte is the president of Althos, an expert information provider covering the communications industry. He has over 29 years of technology analysis, development, implementation, and business management experience. Mr. Harte has worked for leading companies including Ericsson/General Electric, Audiovox/Toshiba and Westinghouse and consulted for hundreds of other companies. Mr. Harte continually researches, analyzes, and tests new communication technologies, applications, and services. He has authored over 30 books on telecommunications technologies on topics including Wireless Mobile, Data Communications, VoIP, Broadband, Prepaid Services, and Communications Billing. Mr. Harte's holds many degrees and certificates include an Executive MBA from Wake Forest University (1995) and a BSET from the University of the State of New York, (1990).

Copyright ©, 2004, ALTHOS, Inc

Table of Contents

Introduction to Bluetooth

Bluetooth is a standardized technology that is used to create temporary (ad-hoc) short-range wireless communication systems. These Bluetooth wireless personal area networks (WPAN) are used to connect personal accessories such as headsets, keyboards, and portable devices to communications equipment and networks.

Bluetooth was named after Harald Blatand, King of Denmark. King Blatand was head of Denmark from 940 to 985 A.D and he is known for uniting the Danes and Norweigans. It seems appropriate to name the wireless technology that unifies communication between diverse sets of devices after King Blatand.

Figure 1.1 shows the different types of devices that can be linked by wireless personal area network communication. This example shows that the computer can be located near the devices such as a keyboard, mouse, display, speakers, microphone, and a presentation projector. As these devices are brought within a few feet of each other, they automatically discover the availability and capabilities of other devices. If these devices have been setup to allow communication with other devices, the user will be able to use these devices as if they were directly connected with each other. As the devices are removed from the area or turned off, the option to use these devices will be disabled from the user.

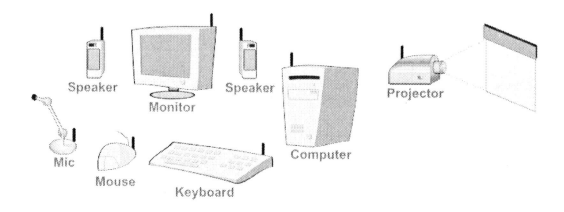

Figure 1.1., Wireless Personal Area Network Devices

The Bluetooth system operates in 2.4 GHz unlicensed (uncontrolled) frequency bands with very low radio transmission power of 1 milliwatt to 100 milliwatts.

For unlicensed use, radio transmission is authorized for all users provided the radio equipment conforms to unlicensed requirements. Anyone can use the unlicensed frequency band but there is no guarantee they will perform at peak performance due to possible interference. Devices within the frequency bands are required to operate in such a way that they can co-exist in the same area with each other with minimal interference with each other.

While users are not required to obtain a license to use devices that operate in unlicensed frequency bands, the manufactures of devices are required to conform to government regulations. These regulations also vary from country to country.

2

Development Timeline

Bluetooth evolved from simple replacement of wires to a dynamically changing wireless personal area network (WPAN). Bluetooth was first conceived in 1993 at Ericsson as a way to allow portable cellular telephones to get smaller while user devices such as PDAs and Laptops could interface with communication devices without the need for wires.

In 1988, several companies setup agreements to form the special interest group (SIG) to develop and promote Bluetooth technology. In July 1999, version 1.0 of the Bluetooth specification was published. In December 1999, additional promoter companies were added and a revised specification 1.0B was released. In 2001, Bluetooth specification 1.1 was released and in November 2003, Bluetooth specification 1.2 was released.

Figure 1.2 shows how Bluetooth evolved from the time it was first conceived in 1993 to its status at the beginning of 2004. This diagram shows that the Bluetooth SIG was formed in 1998 and that version 1.0 Bluetooth specification was released in mid-1999. The first Bluetooth products were qualified in June 2000. There were 83 qualified Bluetooth products by the end of 2000. In February 2001, Bluetooth specification version 1.1 was released and there were 481 qualified products by the end of 2001. An additional 412 Bluetooth products were qualified in 2002 brining the total number of Qualified Bluetooth products to 893 at the end of 2002. In November 2003, Bluetooth specification 1.2 was released and the total number of Bluetooth qualified products that were available at the end of 2003 was 1336.

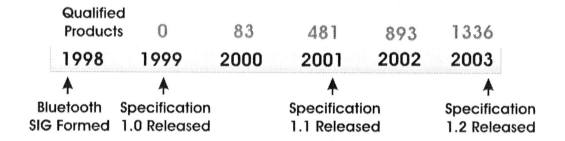

Figure 1.2., Bluetooth_Timeline

Special Interest Group (SIG)

A special interest group works to help develop and promote information about a specific technology, product, or service. The Bluetooth SIG oversees the certification of Bluetooth assemblies and devices and promotes the use of Bluetooth technology.

To allow the SIG to achieve its objectives, the SIG has setup or recognizes specific groups or facilities to be part of the development process. These include the Bluetooth qualification body (BQB), Bluetooth qualified test facility (BQTF), Bluetooth qualification review board (BQRB), and the Bluetooth qualification administrator (BQA).

The BQB is authorized by the Bluetooth qualification review board (BQRB) to be responsible for the checking of declarations and documents against requirements, reviewing product test reports, and listing conforming products in the official database of Bluetooth qualified products. BQTFs are test labs that are recognized and certified by the BQRB as being capable of testing and qualifying Bluetooth devices. The BQRB is responsible for managing, reviewing, and improving the Bluetooth qualification program through which vendor products are tested for conformance. The Bluetooth qualification administrator (BQA) is responsible for overseeing the administration of the qualification program. The BQA ensures that qualified products can be listed on the Bluetooth website.

In addition to the test programs, testing events ("UnplugFests") are made available to members to allow compatibility testing. UnplugFests (UPF) is testing events where manufacturers or developers agree to test their products with other products in a secret closed environment. There are approximately 3 UnplugFests per year. The participation in UnplugFests allows manufacturers and developers to find problems with their products or areas of correction or clarification that are needed in the Bluetooth specifications.

Figure 1.3 shows the general product qualification process used by the Bluetooth special interest group (SIG) to ensure reliable operation and compatibility between Bluetooth devices. This example shows that the first step for product qualification is for the company to gather the program reference documents (PRDs), completes the product documentation, and submit the documents to the Bluetooth Qualification Board (BQB). The company then develops the prototype of the product and submits the product to a Bluetooth Qualification Test Facility (BQTF) for testing. The BQTF test report is then sent to the Bluetooth Qualification Board (BQB) for review. If the product documentation and test results are accepted by the BQB, the product will be added to the Bluetooth qualified product list.

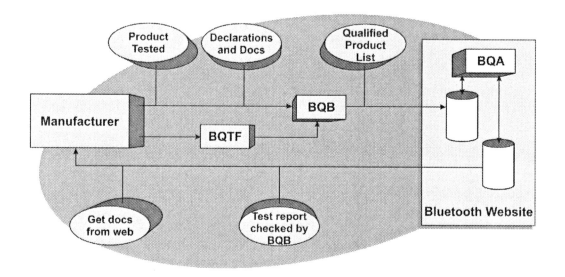

Figure 1.3., Bluetooth Product Qualification Process

Market Growth

Until the late 1990s, the market growth of standardized short-range wireless devices has been relatively low. However, the market for unique proprietary wireless devices such as RF ID tags such auto tollbooth passes (e.g. "E-Zpass"), key chain remotes for cars, remote control garage door openers, two-way radios, and many others has exceeded billions of units. By combining the ability to offer simplicity, mobility, standardization to local area communication systems, low cost wireless devices will continue to grow.

While these wireless PAN devices have existed in proprietary form for tens of years, the ability to use a standard low cost radio module that has advanced feature and security benefits will lead to the conversion of proprietary systems to standard systems. In addition, the availability of low cost radio assemblies will likely lead to the use of wireless in other types of personal area devices that have traditionally been wired devices. These include digital cameras, headsets, keyboards, computer display devices, and other human interface devices (HID).

Another reason for conversion of cables to wireless is the smaller size and lower cost of products. Bluetooth modules are already lower than $5 each (source: Texas Instruments in 2002). The cost of a USB connector, cable cost, and lower market value due to increased product size is well over the $5 cost to insert the module into an electronic device.

Figure 1.4 shows the projected growth of the Bluetooth market. This diagram shows dramatic growth as existing proprietary radio devices convert to a standard Bluetooth radio transmission format and as Bluetooth radios find their way into products with new applications. The total number of

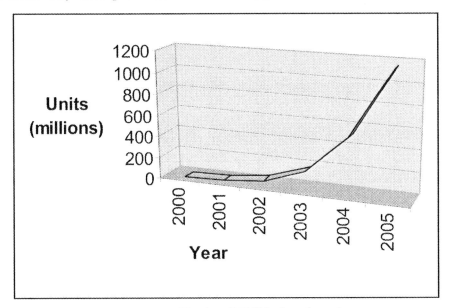

Figure 1.4., Wireless PAN Market Growth
Source: Micrologic Research

Bluetooth products shipped in 2000 was approximately 400,000 units. In 2002, the number of devices shipped with Bluetooth was over 35 million units.

Bluetooth Basics

Bluetooth is a wireless personal area network (WPAN) communication system standard that allows for wireless data connections to be dynamically added and removed between nearby devices. Each Bluetooth wireless network can contain up to 8 active devices and is called a Piconet. Piconets can be linked to each other (overlap) to form larger area Scatternets.

The system control for Bluetooth requires one device to operate as the coordinating device (a master) and all the other devices are slaves. This is very similar to the structure of a universal serial bus (USB) system that is commonly used in personal computers and devices such as digital cameras. However, unlike USB connections, most Bluetooth devices can operate as either a master (coordinator) or slave (follower) and Bluetooth devices can reverse their roles if necessary.

The characteristics of Bluetooth include an unlicensed frequency band that ranges from 2.4 GHz to 2.483 GHz. This frequency band was chosen because it is available for use in most countries throughout the world. While the standard frequency band for Bluetooth is in the 2.400 GHz to 2.483 GHz (83 MHz) frequency band, the original Bluetooth specification had an optional smaller frequency band 23 MHz version for use in some countries. The use of a smaller frequency band does not change the data transmission rate, however, these devices will be more sensitive to interference (such as other Bluetooth device transmission) and this interference may cause a lower overall data transmission rate.

Every Bluetooth device has a unique 48-bit address BD_ADDR (pronounced "B-D-Adder"). In addition to identifying each Bluetooth device, this address is used to determine the frequency hopping pattern that is used by the Bluetooth device.

Bluetooth devices may have different power classification levels. The 3 power versions for Bluetooth include; 1 mW (class 3), 2.5 mW (class 2) and 100 mWatts (class 1). Devices that have an extremely low power level of 1 milliwatt have a very short range of approximately 1 meter. Bluetooth devices have a power level of up to 100 milliwatts can provide a transmission range of approximately to 100 meters. The high power version (class 1) is required to use adjustable (dynamic) power control that automatically is reduced when enough signal strength is available between Bluetooth devices.

Because on of the objectives of Bluetooth is low power and low complexity, the simple modulation type of Gaussian frequency shift keying (GFSK) is used. This modulation technology represents a logical 1 or 0 with a shift of 115 kHz above or below the carrier signal. The data transmission rate of the RF channel is 1 Mbps.

The smallest packet size in the Bluetooth system is the Bluetooth packet data unit (PDU). Bluetooth PDUs are transmitted between master and slave devices within a Bluetooth Piconet. Each PDU contains the address code of the Piconet, device identifier, and a payload of data. When the PDUs are used to carry logical channels, part of the data payload includes a header, which includes logical channel identifiers. The length of the PDU can vary to fit within 1, 3 or 5 time slot period (625 usec per time slot). Control message PDUs (e.g. link control) always fit within 1 time slot.

Figure 1.5 shows that the structure of a Bluetooth contains the address code of the piconet (local system), device identifier (specific device within the piconet), logical channel identifiers (to identify ports), and a payload of data. If a specific protocol is used (such as a wireless RS-232 communication port - RFCOMM), an additional protocol service multiplexer (PSM) field is included at the beginning of the payload data. This diagram also shows that the PDU size can have a 1, 3 or 5 slot length.

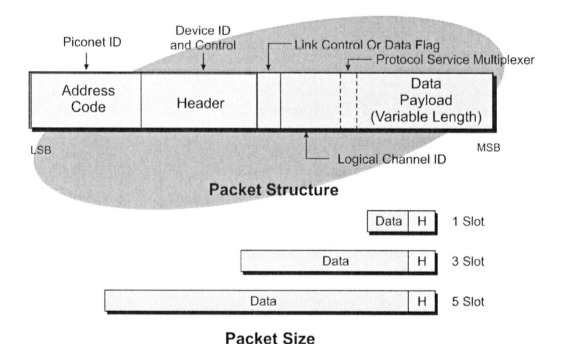

Packet Structure

Packet Size

Figure 1.5., Bluetooth Packet Structure

The Bluetooth system uses time division duplex (TDD) operation. TDD operation permits devices to transmit in either direction, but not at the same time.

Figure 1.6 shows the basic radio transmission process used in the Bluetooth system. This diagram shows that the frequency range of the Bluetooth system ranges from 2.4 GHz to 2.483 GHz and that the basic radio transmission packet time slot is 625 usec. It also shows that one device in a Bluetooth piconet is the master (controller) and other devices are slaves to the master. Each radio packet contains a local area piconet ID, device ID, and logical channel identifier. This diagram also shows that the hopping sequence is normally determined by the master's Bluetooth device address. However, when a device is not under control of the master, it does not know what hopping sequence to use to it listens for inquiries on a standard hopping sequence and then listens for pages using its own Bluetooth device address.

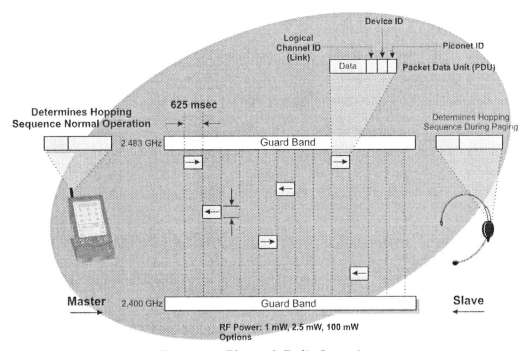

Figure 1.6., Bluetooth Radio Operation

Temporary Small Networks (Piconets)

Bluetooth forms temporary small networks of Bluetooth communication devices of up to 8 active devices called Piconets. The Bluetooth system allows for wireless data connections within the Piconet to be dynamically added and removed between nearby devices. Because the Bluetooth system hops over 79 channels, the probability of interfering (overlapping) with another Bluetooth system is less than 1.5%. This allows several Bluetooth Piconets can operate in the same area at the same time with minimal interference.

Bluetooth communication always designates on of the Bluetooth devices as a main controlling unit (called the master unit). This allows Bluetooth system to be non-contention based. This means that after a Bluetooth device has been added to the temporary network (the Piconet), each device is assigned specific time period to transmit and they do not collide or overlap with other units operating within the same Piconet.

Multiple Piconets can be linked to each other to form Scatternets. Scatternets allow the master in one Piconet to operate as a slave in another Piconet. While this allows Bluetooth devices in one Piconet to communicate with devices in another Piconet (cross-Piconet communication), the use of Scatternets require synchronization (and sharing of data transmission Bandwidth) making them inefficient.

Data Transmission Rates

The basic (gross) radio channel data transmission rate for a single Bluetooth radio channel is 1 Mbps with over 723.2 kbps available to a single user. The data rate available to each user is less than the radio channel data transmission rate because some of the data transmission is used for control and channel management purposes. The users in each Piconet split the remaining data transmission rate.

An example of how the data transmission rate is shared, a Bluetooth Piconet that provides for headset operation, which uses 64 kbps channels in both directions, uses a total data transmission rate of 128 kbps. This is approximately 25% of the total available data transmission bandwidth.

The Bluetooth system allows for different rates in different directions (asynchronous) or for equal data rate (symmetrical rate) transmission.

Figure 1.7 shows how the radio channel data transmission rate for Bluetooth devices is divided between transmission directions and between multiple devices. In example 1, a PDA is transferring a large file to a laptop computer using asymmetrical transmission. During the transfer, it uses the 5-slot packet size to reach the maximum data transmission rate of 723.2 kbps from the PDA to the laptop. This only allows a data transmission rate of 57.6 kbps from the laptop to the PDA. Example 2 shows a symmetrical data transmission rate of 433.9 kbps between two video conferencing stations. Example 3 shows how the data transmission rate from a laptop is shared between a wireless headset and a PDA. This example shows that the headset uses a symmetrical data transmission rate is 64 kbps from device 1 (the master coordinator) to device 2 and a 57.6 kbps asynchronous data transmission rate between the Laptop and the PDA.

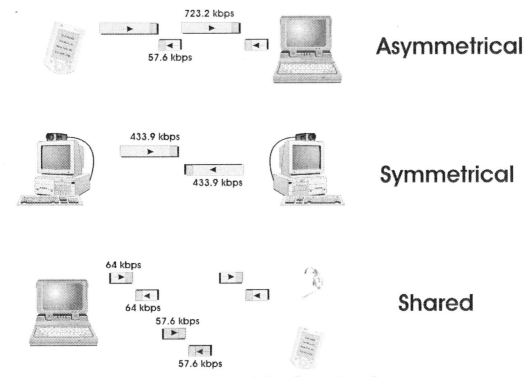

Figure 1.7., Bluetooth Data Transmission Rates

Technologies

The key technologies used in WPAN systems include frequency hopping spread spectrum (FHSS), service discovery protocol (SDP), and established application protocols.

Frequency Hopping Spread Spectrum (FHSS)

Frequency hopping spread spectrum (FHSS) is a radio transmission process where a message or voice communications is sent on a radio channel that regularly changes frequency (hops) according to a predetermined code. The receiver of the message or voice information must also receive on the same frequencies using the same frequency hopping sequence. Frequency hopping was first used for military electronic countermeasures. Because radio communication occurs only for brief periods on a radio channel and the frequency hop channel numbers are only known to authorized receivers of the information, transmitted signals that use frequency hopping are difficult to detect and monitor.

Figure 1.8 shows a simplified diagram of how the Bluetooth system uses frequency hopping to transfers information (data) from a transmitter to a receiver using 79 communication channels. This diagram shows a transmitter that has a preprogrammed frequency tuning sequence and this frequency sequence occurs by hopping from channel frequency to channel frequency. To receive information from the transmitter, the receiver uses the exact same hopping sequence is used. When the transmitter and receiver frequency hopping sequences occur exactly at the same time, information can transfer from the transmitter to the receiver. This diagram shows that after the transmitter hops to a new frequency, it transmits a burst of information (packet of data). Because the receiver hops to the same frequency, it can receive the packet of data each time.

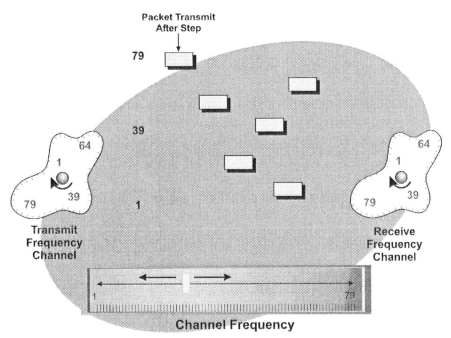

Figure 1.8., Bluetooth Frequency Hopping Operation

Interference with Other Devices

Bluetooth operates within the same frequency band, as other devices so there will be interference to and from other devices. Some of common types of other 2.4 GHz devices that Bluetooth interferes with include cordless telephones, microwave ovens, wireless cameras, and 802.11 wireless LAN.

Some 2.4 GHz cordless telephones use simple FM modulation. These cordless telephones have a relatively narrow bandwidth so only a small portion of the Bluetooth radio channels will be affected. Inexpensive (residential) microwave ovens only transmit for short cycles of 16 to 20 msec on and 16 to 20 msec off. This interference will only occur when the microwave is operating and when Bluetooth is transmitting, the interference is relatively small. Wireless cameras (such as a security monitoring system) may continuously transmit and can interfere with 6-8 Bluetooth RF channels. One of the most difficult interference sources is 802.11 Wireless LAN (WLAN). The most popular version of 802.11 (802.11b) uses 25 MHz wide channels that are very sensitive to radio distortion. This means that 802.11b is more sensitive to Bluetooth than Bluetooth is sensitive to 802.11.

Because Bluetooth transmits over a wide frequency bandwidth of 79 channels, much of the Bluetooth packets will get through even the toughest interference sources. Because Bluetooth transmits at such a low power, the amount of interference Bluetooth causes to other systems is low.

Figure 1.9 shows the some of the types of devices that interact with Bluetooth and how they may interfere with each other. This diagram shows that a 2.4 GHz cordless telephone only uses a relatively narrow radio channel and it only interferes with a fairly small amount of Bluetooth packets. The microwave interferes with a larger number of packets, yet the burst nature of microwave ovens still allow many Bluetooth packets to get through. The interference of a wireless security camera is constant and can interfere with a small percentage of Bluetooth packets. This example shows

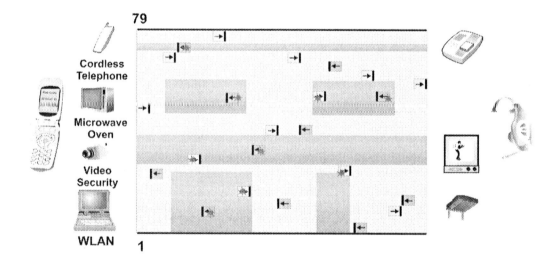

Figure 1.9., Bluetooth Radio Interference

16

that the most significant interference occurs with 802.11 WLAN systems because it uses a relatively high-power 25 MHz radio channel that is also sensitive to even small amounts of radio distortion.

Adaptive Frequency Hopping

Bluetooth specification 1.2 introduced adaptive frequency hopping (AFH) that can reduce the effects of interference between Bluetooth and other types of devices. The AFH adapts the access channel sharing method so that the transmission does not occur on channels that have significant interference. By using interference avoidance, devices that operate within the same frequency band and within the same physical area can detect the presence of each other and adjust their communication system to reduce the amount of overlap (interference) caused by each other.

The adaptive frequency hopping process reassigns the transmission of packets on channels that have interference to other channels that have lesser interference levels. This reduced level of interference increases the amount of successful transmissions therefore increasing the overall efficiency and increased overall data transmission rate for the Bluetooth device and reduces the effects of interference from the Bluetooth transmitter to other devices.

Figure 1.10 shows how the Bluetooth specification 1.2 allows the Bluetooth device to change its hopping pattern to avoid interference to and from other devices that operate within its frequency band. This example shows that a video camera that is using multiple frequency channels. After detecting the presence of a continuous signal being transmitted by the video camera in the 2.4 GHz frequency band. The Bluetooth device automatically changes its frequency hopping patter to avoid transmitting on the frequency band that is used by the video camera signal transmission. This results in more packets being successfully sent by the Bluetooth device and reduced interference from the Bluetooth device to the transmitted video signal.

79

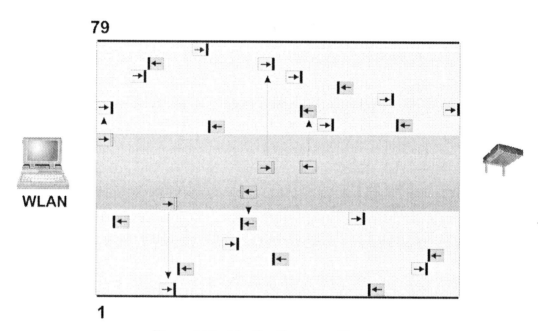

1

Figure 1.10., Adaptive Frequency Hopping

Service Discovery

Service discovery is the process of finding other devices that can communicate with your device and determining what capabilities they have that you may want to use. Service discovery protocol (SDP) is the communication messaging protocol used by a communication system (such as the Bluetooth system) to allow devices to discover the availability and capabilities of other nearby devices. The SDP process is similar to a registry in Windows as it dynamically creates a list of available resources.

The discovery process begins with an inquiry message that a device sends that can be received by nearby devices. These devices are constantly looking for an inquiry message to respond to. When a device receives an inquiry message, it responds with this address that can be used to establish a connection with the device.

If a device wants to discover the services of another device, it must use the device address and establish a temporary connection. The name and capabilities of the device can be discovered using service discovery protocol (SDP). The discovery process is optional. Devices can be programmed not to respond to inquiry messages.

Figure 1.11 shows the typical service discovery operation of a PDA device that desires to print to a nearby printer. In this example, a PDA unit (acting as a master unit) sends out an inquiry message to a printer (a slave unit). The printer (and possibly other devices) responds with its' Bluetooth device address. The PDA then sends a connection request using the printer's Bluetooth address along with a request for the capabilities of the printer. The printer then returns the capabilities requested (if available) from its SDP database and these capabilities are temporarily stored in the SDP database of the PDA. The PDA can now display the availability of the printer to the user. This allows the user to select the printer of choice that has specific printing capabilities (such as laser or color printing).

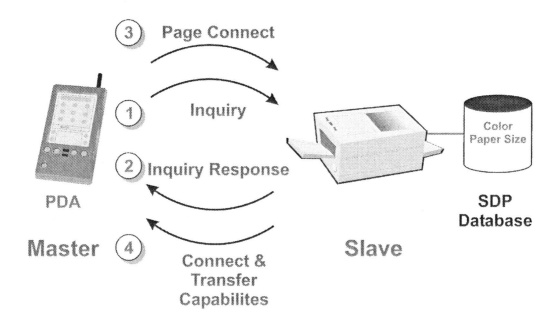

Figure 1.11., Service Discovery Protocol

Application Protocols

Application protocols are commands and procedures used by software programs to perform operations using information or messages that are received from or sent to other sources (such as a user at a keyboard). Application protocols are independent of the underlying technologies and communication protocols. The use of well-defined application protocols (agreed commands and processes) allows the software applications to interoperate with other programs that use the application protocol independent from the underlying technologies that link them together (such as wires or wireless connections.)

When Bluetooth system was developed, there were already many standard application protocols already available. The Bluetooth system used or adapted these protocols to allow applications to communicate via Bluetooth radio. Some of the popular application protocols that are used include RS-232 serial data connection, point-to-point protocol (PPP), object exchange (OBEX), and telephone call control protocol Q.931.

Figure 1.12 shows how the Bluetooth system uses standard industry protocols to connect to standard communication applications. In this example, a laptop computer is communicating with a wireless mouse, a personal digital assistant (PDA), and an access node using a single WPAN PCMCIA card. When communicating with the mouse, the laptop uses the standard RS-232 protocol. When transferring (exchanging) items between an address book stored in the laptop and an address book stored in the PDA, it uses the standard Object Exchange (OBEX) protocol. To connect to the Internet, the laptop connects through an access node to a router using standard point-to-point protocol (PPP). This diagram shows that the PPP connection is only part of the communication link that reaches an email server that is connected to the Internet. The computer uses standard simple mail transfer protocol (SMTP) to send and retrieve email messages.

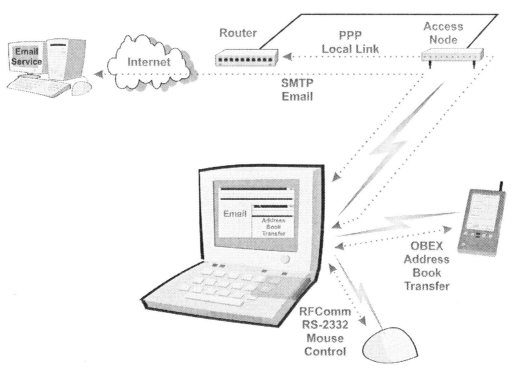

Figure 1.12., Application Protocols

Bluetooth Operation

Bluetooth operation involves looking for other devices (called Inquiring), creating a secure (linked) relationship with these devices (called pairing and bonding), connecting with the devices to transmit data, and setting up communication sessions between devices.

Inquiring and Finding Other Bluetooth Devices

Inquiry is a process in a Bluetooth system that is used to determine the address and name of other Bluetooth devices that are operating in the same area. Bluetooth inquiry is the process that requests specific information from a computer or communication device to determine its access code or availability for a communication session.

The discovery process can take up to 10 seconds and the ability to allow a device to be "discoverable" is optional. If a user does not want a device to be discovered, the device can be placed in "non-discoverable" mode. Some devices (such as mobile telephones) come programmed to be non-discoverable.

Figure 1.13 shows a simplified diagram of how the Bluetooth system inquiry process discovers nearby Bluetooth communication devices. This diagram shows each Bluetooth transmitter and receiver has a standard preprogrammed frequency tuning sequence called the general inquiry access code (GIAC). The GIAC access code causes the transmitter and receiver frequency sequence to change from one channel frequency to channel frequency. Because the transmitter and receiver do not know exactly when each will transmit, to initially capture information from a transmitter to a receiver, the transmitter changes frequency often (3200 steps per second). The receiver changes frequency using the same hopping sequence, but at a much slower hop rate (possibly 1 step every 1.28 seconds). Each time the transmitter stops on a specific frequency, it transmits two short inquiry identification packets (they contain the GIAC access code). Eventually, the transmitter and receiver wind up on the same channel and the receiver will capture at least one of these inquiry packets. After waiting a random amount of time to avoid collisions with other Bluetooth devices that may have also received the inquiry packet, the receiver will respond to the transmitter informing it of its address. The transmitter has now discovered that another device is operating near it and it can use that address to contact (page) the other Bluetooth device and setup a communication session.

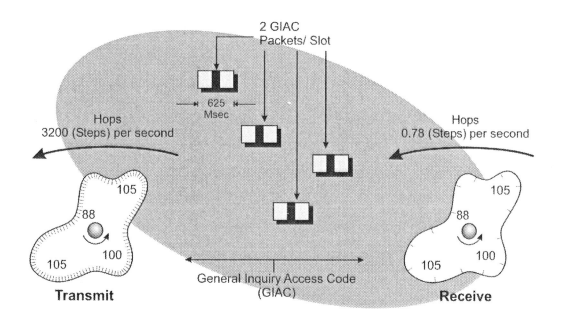

2 GIAC
Packets/ Slot

625
Msec

Hops
3200 (Steps) per second

Hops
0.78 (Steps) per second

105

88

105   100

Transmit

General Inquiry Access Code
(GIAC)

105

88

105   100

Receive

Figure 1.13., Bluetooth Inquiry Operation

Pairing with Other Bluetooth Devices

Bluetooth pairing is an initialization procedure whereby two devices communicating for the first time create an initial secret link key that will be used for subsequent authentication. For first-time connection, pairing requires the user to enter a Bluetooth security code or PIN.

Since neither unit knows any secret keys of the other unit, a new secret key must be created. As a result, both devices request a PIN code to be entered. Both users must enter the same PIN code. This PIN code is used with other information to create a secret key in each unit

The Bluetooth pairing process allows devices to authenticate and create a secure link between two Bluetooth devices. For example, if user A desires to push a business card to user B, user A attempts to establish a connection with device B. If device B has been setup to reject the connection unless a device has been paired, device B requests authentication of device A. This initiates a screen or alert on both devices requesting the entry of a PIN code. Both users must enter the same PIN code. This PIN code is combined with other information to produce a secret key. Assuming the same PIN code is entered by both users, the secret key that is created is the same for both users and this key can be used to help authenticate (validate) the identify of the other user.

Bonding with Other Bluetooth Devices

Bonding is the process of creating a very secure link key that is shared between Bluetooth devices. One way to create a very secure link key is the use the initial secret link key that was created by the pairing process. The bonding process involves encoding (modifying) a unique unit key (created when the Bluetooth device is turned on) using the initial secret link key and sending the unit key to the other unit. This allows the Bluetooth devices to use each other's secret information to create a more secure link key. This secure link key can be used in future authentication validations.

Connecting with Other Bluetooth Devices

Connecting Bluetooth devices is the process of creating a communication session between devices. Communication sessions are the end-to-end transmission links between devices during operation of a software program or logical connection between two communications devices. In communications systems, the session involves the establishment of a physical channel, logical channel(s), and the configuration of transmission parameters, operation of higher-level applications, and termination of the session when the application is complete. During a session, many processes or message transmissions may occur.

Creating a connection between Bluetooth devices involves getting the attention of a device through paging and allowing the device to change it's frequency hopping sequence so that it can become part of the Piconet.

If the Bluetooth device knows the address of the Bluetooth device it wants to connect to, the connection process usually takes less than 1-2 seconds. The ability of a device to allow "connections" is optional. Bluetooth devices can be programmed to not allow other devices to connect to them ("non-connectable.")

The connection process begins with the master unit changing its hopping sequence to the hopping sequence of the recipient device. The Bluetooth unit first sends many identification (ID) packets alerting the receiving device that someone wants to connect to them. When the receiving device (the slave) hears its ID address, it can immediately respond as no other units will be competing for its own access code. When the master hears that the recipient has responded, it sends a frequency hopping synchronization (FHS) packet that contains its Bluetooth Address. Both the master and slave then change their hopping sequence to the master's hopping sequence (the Piconet address). The master will then send a Poll message to the slave using the new hopping sequence and if the slave responds (usually with a null-no information response), the master knows it is connected to the recipient (slave) device.

Figure 1.14 shows how a Bluetooth device can connect to other devices. This example shows that the Bluetooth master unit first sends many ID packets using the hopping sequence of the recipient device. When the receiving device hears its ID address, it immediately responds. This allows the master to send a FHS packet that contains the masters Bluetooth Address. Now both the master and slave change their hopping sequence to the Piconet hopping sequence (determined by the master's Bluetooth Address). The master then sends out a Poll message using the new hopping sequence and the slave will typically respond with a null response.

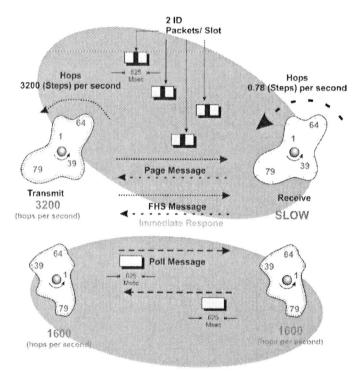

Figure 1.14., Bluetooth Connection Operation

Bluetooth Sleep Modes

Bluetooth sleep modes include short one-time hold periods, periodic sniff periods, and long time park periods. Bluetooth sleep modes are used to reduce the power consumption (extend the battery life) and to free the Piconet of device activity so other devices may participate in the Piconet.

Hold is a temporary mode of operation that is typically entered into by a device when there is no need to send voice or data information for a relatively long time. The hold mode allows the device audio to be muted or the transceiver to be turned off in order to save power. The Bluetooth hold mode is used to release devices from actively communicating with the master. This allows the devices to sleep for short periods and allows the master control device to discover or be discovered by other Bluetooth devices that want to join other Piconets. With the hold mode, Piconet capacity can be freed up to do other things like scanning, paging, inquiring, or attending another Piconet sessions.

Sniffing is a process of listening for specific types of commands that occur periodically. Sniffing is used for devices that must continuously be in contact with the master. The Bluetooth sniff mode is used to reduce the power consumption of the devices as the receiver can be put into standby between sniff cycles.

Park is the process of temporarily deactivating a device to allow its active member address to be removed (probably re-assigned) and assigning communication functions to remain inactive for extended periods of time. When the master commands the slave to part, the slave will periodically wake up and look for a beacon signal from the master unit. If this beacon signal contains the addresses of the parked device, the device will reactive and become part of the Piconet again. The maximum time period that can be assigned for hold, sniff, or park sleep mode is 65,440 slots (approximately 40 seconds).

Figure 1.15 shows the different types of sleep modes used in a Bluetooth system. This diagram shows that the master control the sleep modes of the devices within the Piconet. In this example, the master in the Piconet commands a PDA to one time sleep period (hold mode) for 500 msec. The master then commands the mouse in the Piconet to sleep periodically (sniff mode) for 50 msec. The master then commands the scanner to give up its active member address wait until 10 seconds before it will again communicate with the scanner.

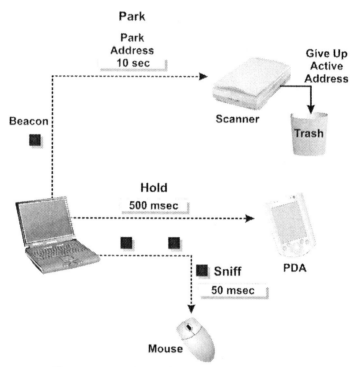

Figure 1.15., Bluetooth Sleep Mode Operation

Protocols

Protocols are a precise set of rules, timing, and a syntax that govern the accurate transfer of information between devices or software applications. Key protocols in data transmission networks include access protocols, handshaking, transmission parameter negotiation, and session protocols. The Bluetooth system consists of many existing protocols that are directly used or have been adapted to the specific use of the Bluetooth system.

Protocols are often divided into groups that are used for different levels of communication. Lower level protocols (such as protocols that are used to manage a radio link between specific points) are only used to create, manage, and disconnect transmission between specific points. Mid-level protocols (such as transmission control protocols) are used to create, manage, and disconnect a logical connection between endpoints that may have multiple link connections between them. High level protocols (application layer protocols) are used to launch, control, and close end-user applications.

Some of the important protocols associated with the Bluetooth system include the host controller interface (HCI), logical link control applications protocol (L2CAP), service discovery protocol (SDP), RF Communications protocol (RFCOMM), Object Exchange (OBEX), and Telephony Control Protocol (TCS).

Host Controller Interface (HCI)

Host controller interface (HCI) protocol is used to directly control a Bluetooth radio assembly (module or device) through a serial data connection. Use of the HCI protocol is optional. However, if the HCI protocol is used, the commands, data transfer, and electrical connections are precisely defined so a radio module created by one manufacturer can be directly exchanged with a radio module from another manufacturer.

The HCI protocol is composed of a set of commands and events and their associated parameters. The commands are used to setup and control the Bluetooth radio. Events are the messages received by the Bluetooth radio that require the attention of the host application.

The HCI protocol allows direct control of the radio module and how a radio module from one manufacturer can be directly exchanged with a radio module from another manufacturer. For example, this allows a laptop computer to operate as the host controller of a Bluetooth radio using an HCI RS-232 data connection. The host application on the laptop computer sends control commands and data messages to control the radio via the RS-232 serial data line. The Bluetooth radio sends event information and received data packets to the host program on the laptop computer.

Logical Link Control Applications Protocol (L2CAP)

Logical link control application (L2CAP) protocol manages the logical connections between the application layers and the physical layer in a Bluetooth communication system. This protocol performs logical channel multiplexing (managing multiple logical channel connections), packet segmentation and re-assembly (SAR), and group link management. The L2CAP allow applications to transmit and receive data packets of up to 64 Kbps in length.

The L2CAP protocol is used to multiplex multiple logical channels on a single physical link. A connection identifier identifies each logical channel. At least one logical channel is reserved as a control channel to setup and manage the logical connections.

Figure 1.16 shows how the L2CAP protocol can create, manage, and terminate logical connections between Bluetooth devices. This diagram shows that the L2CAP receives and sends packets to applications through the use of logical channels and each logical channel is identified by a connection identifier (CID). This example shows that the L2CAP protocol also combines (assembles) and divides (disassembles) the packets to and from the applications into smaller packets that can be communicated to the Bluetooth radio module.

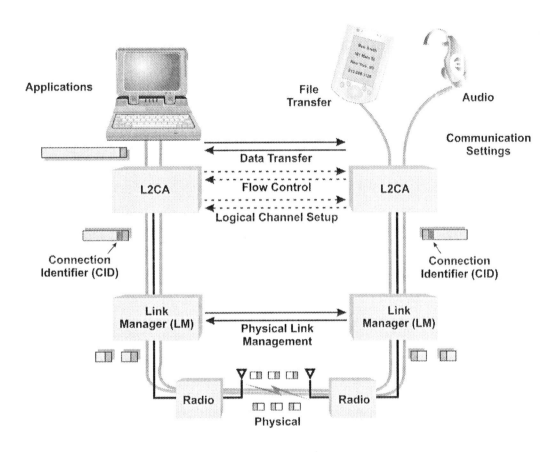

Figure 1.16., Logical Link Control Application Protocol Operation

Service Discovery Protocol (SDP)

Service discovery protocol is used to allow devices to discover the services and capabilities of other devices. Bluetooth devices use SDP to setup a communication session that allows devices to generally and/or specifically search for capabilities and services of other communication devices.

SDP messages are used to setup and inquire into service records of devices. Service records contain the information that is necessary to identify and use the service of the device. A record handle identifies each service record. Once the inquiring device has obtained the service handle, it can access additional details (attributes) of the service.

The service capabilities and records of a device may change during the device. As devices are connected and removed from the Bluetooth capable device, the appropriate service records are added or removed.

Bluetooth devices use SDP protocol to identify and transfer service information between other devices. For example, a PDA can send an SDP message to a computer requesting if it has printer capability. When the computer receives the SDP message, it searches into the SDP Registrations to find the printer record and it returns an SDP message with the record handle identifier.

RF Communications Protocol (RFComm)

RF Communications (RFComm) protocol is to emulate the operation of a serial communication port. This protocol allows for the setup of serial communication channels and the simulation of hardware pins RS-232 physical connections. The RFComm protocol emulates the software and hardware operations of RS-232 (EIA/TIA-232-E) serial ports and is based on ETSI specification TS 07.10.

The RS-232 serial communications specification is an electronics industry association (EIA) standard protocol that is used to transfer information in asynchronous (unscheduled) form. The RS-232 specification defines optional physical (connector), electrical (signal levels), and software data formats. RS-232 serial connectors are a common connector that is used on the back of computers to connect to modems and other external devices. RS-232 communication uses universal asynchronous receiver and transmitter (UART) technology and adds communication negotiation features to the serial data communication session.

Figure 1.17 shows the RFComm protocol can be used to emulate a wired connection between a serial mouse and a standard computer RS-232 port. This example shows that the computer requests an RS-2323 connection request to the Bluetooth radio through the RF communication port. The RFComm protocol determines it needs a communication session so it requests that the L2CA layer initiate a logical connection to the mouse Bluetooth radio. Once the logical channel is created between the laptop computer and the Bluetooth mouse, the RFComm protocol is used to establish the parameters (settings) of the serial connection (data rates and connection control type). The RFComm protocol then continually receives serial data from the mouse, identifies (addresses) and packetizes the data, and sends it to the laptop computer.

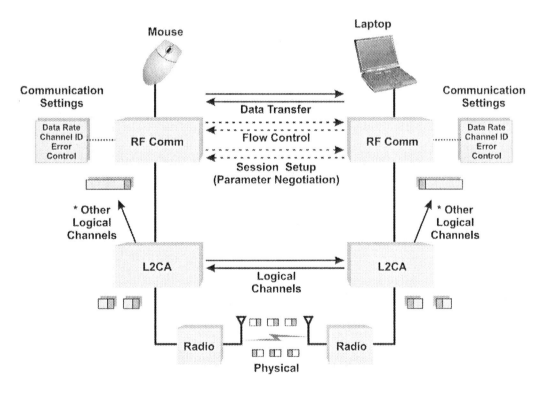

Figure 1.17., RFComm Protocol Operation

Object Exchange Protocol (OBEX)

Object exchange protocol (OBEX) is a session-layer protocol for object exchange originally developed by the Infrared Data Association (IrDA) as IrOBEX . Its purpose is to support the exchange of objects in a simple and spontaneous manner over an infrared or Bluetooth wireless link.

Object exchange is a high-level protocol (management protocol) that is used for object (data) exchange. OBEX was originally developed by the Infrared Data Association (IrDA) as IrOBEX. Its purpose is to support the exchange of objects in a simple and spontaneous manner over an infrared or Bluetooth wireless link.

The first step in the OBEX protocol is to establish a communication session. This involves creating a logical communication path between the devices and determining the version of OBEX protocol that is running on both machines. OBEX protocol defines messages that include the connection and disconnection of Object Exchange sessions and information is requested or sent using Get or Put messages.

Telephony Control Protocol (TCS)

Telephony control protocol is used for the establishment of speech and data calls between Bluetooth devices. In addition, the TCS protocol can be used for mobility management to coordinate groups (multiple telephone handsets) of Bluetooth TCS devices. The Bluetooth TCS protocol uses standard AT commands by which a mobile phone and modem can be controlled.

The reason telephone control protocol is not called TCP to avoid confusion with transmission control protocol (TCP) used for Internet communication.

The TCS protocol is based on the standard Q.931 is a telephone control protocol that was developed for the call control in the public telephone network (ISDN). The Q.931 call control protocol defines the messages and formats of control messages that are created by the end communication devices (such as telephones and fax machines). Some of the common types of information contained in Q.931 messages include call setup and tear down messages, called and calling party telephone numbers, and other access control signaling messages. Because Q.931 protocol has been reliably used in the telephone system for many years, the Bluetooth system adapted much of the Q.931 protocol messages for call processing (setup, connection, and disconnection).

Figure 1.18 shows how the TCS protocol can be used to emulate a telephone connection between a cordless telephones and a cordless telephone base. This example shows that the Bluetooth cordless telephone (terminal) requests a telephone call to the Bluetooth cordless base (gateway). The TCS protocol first determines that it needs a communication session so it

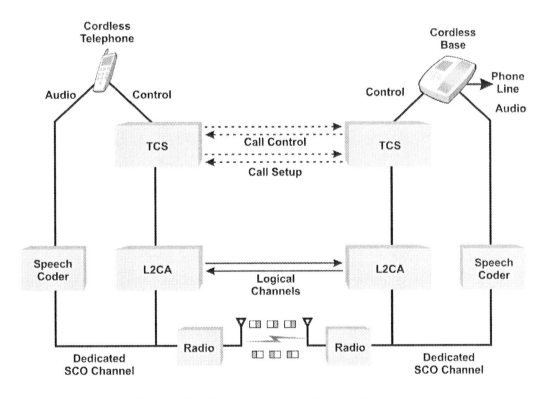

Figure 1.18., Telephony Control Protocol Operation

requests that the L2CA layer initiate a logical connection to cordless base station Bluetooth radio. Once the logical channel is created between the cordless telephone and the cordless base, the TCS protocol is used to setup the call request (transfer the dialed digits). This diagram shows that the TCS protocol also requests that a dedicated audio connection (synchronous channel) through a speech coder. The TCS protocol then continually manages the progress of the call (such as initiating a three-way call) until the call is ended.

Profiles

Bluetooth profiles are particular implementations, processes, and definitions of the required operations and protocols that allow Bluetooth provide a specific service or application to ensure interoperability between Bluetooth devices. Profiles define the required protocols and other supporting profiles (such as how to obtain general access to other Bluetooth devices) that are required to provide types of specific services or features to the user. The conformance to Profiles helps to ensure reliable operation of Bluetooth devices regardless of who manufactured the device or which version of the product is used.

Figure 1.19 shows how profiles are used to standardize how Bluetooth devices communicate with each other independent of who manufactured the device. This diagram shows that profile includes the required communication protocols, the service capabilities, and feature operation of the Bluetooth device. In this example, the Bluetooth profile also defines how the user can interact (features and optional features) with the device and how the device will display information to the user.

Introduction to Bluetooth

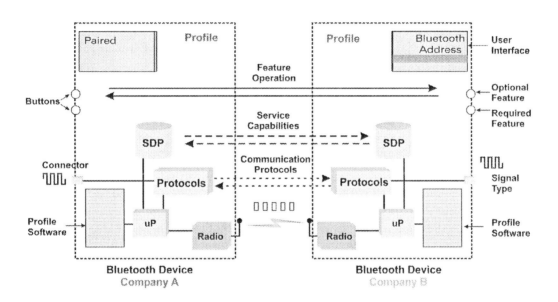

Figure 1.19., Bluetooth Profile Operation

Common Bluetooth profiles general access profile (GAP), service discovery
profile (SDP), general object exchange (GOEP), headset, synchronization,
serial port, human interface device (HID), basic printing, and hands free
profile.

General Access Profile (GAP)

Bluetooth generic access profile describes the use of the lower layers of the
Bluetooth protocol stack-Link Control (LC) and Link Manager Protocol
(LMP) to ensure that devices can reliably discover and connect to each other.
Also defined in this profile are security-related procedures, in which case
higher layers-L2CAP, RFCOMM, and OBEX-come into play.

The general access profile (GAP) is used to help ensure that Bluetooth devices can reliably discover and connect to each other. For example, if a PDA desires to connect to a laptop computer to transfer files, the GAP specifies the amount of time the PDA must transmit inquiry messages for and how often and how long the laptop computer must listen for to ensure the PDA transmission and the laptop receiver eventually communicate with each other.

Service Discovery Application Profile (SDAP)

Service Discovery Application Profile (SDAP) defines the processes and procedures required to allow devices to discover and allow other devices to discover services available for a Bluetooth device. The SDP allows for the dynamic addition and elimination of available services dependent on the capability and status of the Bluetooth device. The use of the SDAP ensures that Bluetooth devices and retrieve information pertinent to the implementation of services as they become available in other Bluetooth devices.

General Object Exchange

Bluetooth general object exchange profile defines the protocols and procedures used by the applications providing the usage models that need object exchange capabilities. Examples of such usage models include Synchronization, File Transfer, and Object Push. The most common devices using these usage models include notebook PCs, PDAs, smart phones, and mobile phones.

General object exchange profile (GOEP) is used to define what objects and how information objects are transferred between communication devices. For example, an address book data file contained in a PDA is divided into blocks of information that have specific formats. The definition of these specific formats allows parts of the address book in the PDA file to be accurately exchanged or updated with other address books that have the same format.

Headset

Bluetooth headset profile defines the processes and options available for headsets to communicate with audio capable devices. The headset profile defines the types of audio connections (types of digital audio) used to link audio devices such as mobile telephones and CD players with a wireless headset.

The use of the headset profile allows a headset that is produced from one manufacturer to be used with a mobile telephone from another manufacturer. To enable headset operation, the headset is first paired with the mobile telephone by pressing specific keys on the mobile telephone. This creates a shared secret key that is used during connection of the headset to the mobile telephone control and audio sections. The headset profile also defines the control options for the headset. In this example, the headset includes a volume control key that allows the user to adjust the volume of the mobile telephone from the wireless headset.

Serial Port Profile

Serial Port Profile provides the required processes and physical connections (if necessary) to allow simple serial data connection that emulates RS-232 serial ports (cable or software) between two peer devices. The serial port profile emulates the setup commands and electrical control pins between two peer Bluetooth devices using RFCOMM protocol.

Synchronization

The synchronization profile defines the processes used to ensure new information that that is entered into one storage system can be reliably exchanged and replaced with information that is stored on another system. The Bluetooth synchronization profiles allows for manual or automatic synchronization. Manual synchronization occurs when a user initiates the synchronization process and automatic synchronization occurs Bluetooth devices are setup to synchronize when they automatically detect the presence (availability) of another Bluetooth device.

Synchronization works by assigning a date and timestamp to information as it is entered into an information file storage system. For example, as a new address is entered into an address book file, this new information has a more current date and timestamp than other addresses stored in that address book. During the synchronization process, the date and time stamp information from two (or more) information files are compared. Information in one file that has a more recent date can be exchanged and replaced with older information in the other file.

Figure 1.20 shows how Bluetooth synchronization can be used to update information that is arbitrarily entered into devices that are able to time stamp and exchange information. This diagram shows that a user has updated their personal digital assistant (PDA) with a new address for a person. Later on, the user updates their mobile telephone with a new phone number for the same person. This example shows that the user has setup the synchronization service with auto-detection capability. When the PDA

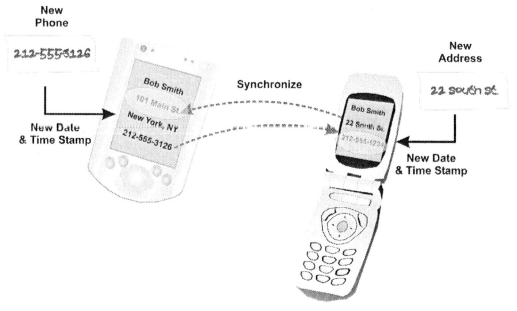

Figure 1.20., Bluetooth Synchronization Profile

senses the mobile telephone is close enough, the PDA and mobile telephone will check for information that has more recent date and timestamps. When the PDA determines that the mobile telephone has a newer phone number, it will automatically transfer and update (replace) the old phone number. When the mobile telephone determines that the PDA has a more recent address, it will automatically transfer and update (replace) the old address.

Human Interface Device (HID)

Bluetooth human interface device (HID) profile is describes the operation and protocol requirements to allow typical human interface devices such as mice, keyboards, and game pads to communicate with the systems they interact with. The HID protocol is based on the USB definition. The HID profile groups devices according to their class and each class has their own drivers. HID devices are smart and they can describe their capabilities (such as what buttons are available on a game pad) to the system. HID devices can be hot swapped allowing them to be added or removed at any time without causing problems to the system.

The use of the human interface device (HID) profile allows various types of standard accessories (such as a wireless keyboard) to be used with a computer or a mobile telephone provided these devices have HID profile capability. For example, a user can connect their mobile telephone to a wireless keyboard to permit more rapid entry of instant messages (SMS).

Basic Printing

Bluetooth printing profile are the required processes and methods used to allow Bluetooth devices to print multiple types of information to printers with different capabilities. The Basic printing profile allows for Bluetooth devices such as mobile telephones and PDAs to identify available printers, gather information from these printers necessary to communicate with them, and to adjust the format information sent to the printer to achieve a printed look.

Figure 1.21 shows how the Bluetooth printer profile can be used to allow a digital camera to find available printers that meet its specific printing requirements. In this example, the digital camera with Bluetooth has discovered 2 printers. By sending a SDP command requesting color capability, printer B responds that it has color capability so the digital camera connects to printer B and sends the picture to B for printing.

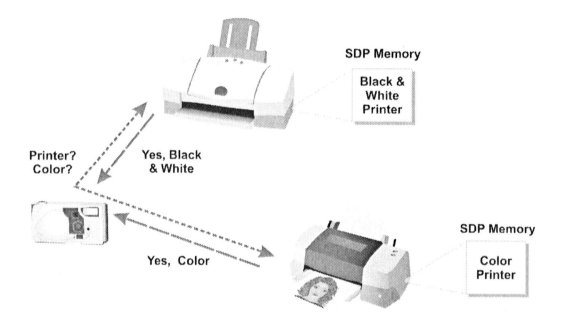

Figure 1.21., Bluetooth Printing Profile

Hands Free (Automotive Car Kit)

Bluetooth hands free profile defines how mobile telephones will interact with automobiles to allow for hands free operation. This profile specifies the hands free kit gateway communicates with the mobile telephone (the audio gateway) that communicates with the cellular telephone network. The hands free kit may include several features such as voice activated dialing, memory dialing, and echo canceling.

Figure 1.22 shows how the Bluetooth Car Kit profile allows a mobile telephone to provide for handsfree operation by interacting with the automobiles audio system. This example shows a mobile telephone that has discovered the car's Bluetooth signal and has established a connection with the car. This connection allows incoming calls to alert the car's audio system that the radio should be muted and the external microphone to be activated.

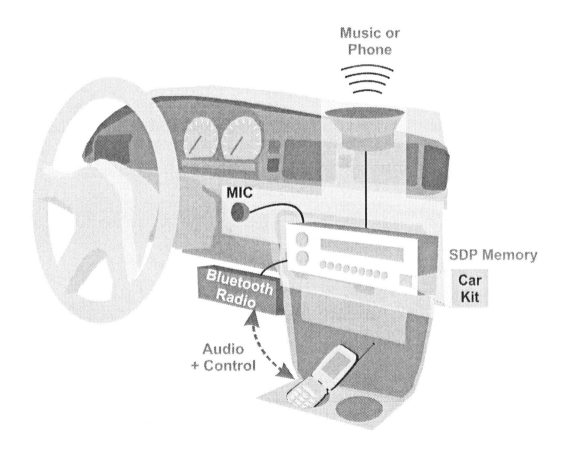

Figure 1.22., Bluetooth Car Kit Profile <ag_Bluetooth_Headset_Profile

Other Profiles

There are many other profiles including cordless telephone, intercom (two-way radio like), LAN Access, general audio and video, fax, dial-up modem, hardcopy cable replacement, and many others.

Intercom Profile defines the protocols and procedures that would be used by Bluetooth devices for implementing the intercom part of the usage model called "3-in-1 phone," also known as the "walkie-talkie" application of the Bluetooth specification.

The LAN access profile defines how the Point-to-Point Protocol (PPP) is used to provide LAN access services for Bluetooth devices.

The General Audio Video Distribution Profile (GAVDP) defines the requirements for sending audio and video streaming signals.

The fax profile describes the protocols or procedures that adapt an information format into a format suitable for facsimile communication.

The Dial-up Modem Profile defines the protocols and procedures to allow data communication devices to use another communication device (such as a cellular telephone) to act as a bridge. This bridge (an "Internet Bridge") can be used to allow a computer to connect to a dialup Internet service provider.

Cordless Telephone Profile describes a "3-in-1 phone" solution; cellular, cordless, and intercom. The 3-in-1 phone profile can be used to integrate personal (residential) phone service with wide area (business) service.

Profiles Updates

Devices Profiles will evolve to include new features and companies will add new profiles to existing devices. There are several ways to get new profiles and updated profiles for your Bluetooth capable device. Getting new profiles may be as simple as downloading a new software version or it may involve having the product serviced. For profiles that are stored in mobile telephones, it may be necessary for the user to visit a service center so a new mobile telephone operating system can be installed.

Future Enhancements

Future enhancements to WPAN systems include much higher data transmission rates, more rapid signal acquisition, and better co-existence through interference avoidance.

Higher Data-Rates

It is likely that more advanced modulation technologies will be used in wireless PAN systems. Receivers that are capable of more precisely demodulating phase and amplitude signals will be able to transmit and receive at higher data transmission rates. Some of the early LAN systems (such as 802.11) transmitted at a data rate of 1 Mbps. These systems evolved through the use of advanced modulation technologies to increase their data rates up to 54 Mbps.

Modulation is the process of transferring the information signals onto a signal that is used for transmission through the use of amplitude changes, frequency changes or phase changes. Complex signal processing allows for advanced digital modulation efficiency. More efficient digital modulation has increased the amount of bits per Hertz of bandwidth that can be sent. Early systems provided approximately 1 bit per Hertz. Present systems provide 5 to 10 bits per Hertz.

Increased data transmission rates through the use of more advanced (more complex) modulation technologies. In general, there is a tradeoff between complex modulation technologies and susceptibility to interference from other devices. For example, a device that represents data by frequency changes is not as susceptible to changes in amplitude caused by lightning.

Advanced forms of modulation often use combined phase and amplitude. While distortions may cause short bursts of errors, the overall increased data transmission rate may provide much higher overall data transmission rates.

Figure 1.23 shows how wireless systems (such as WPAN systems) can increase their data transmission rates through the use of advanced modulation technologies. The first WPAN systems used simple frequency or amplitude modulation to achieve data transmission rates of about 1 to 2 bits per Hertz of available bandwidth. More advanced WPAN systems combine multiple forms of modulation to achieve data rates in excess of 2 to 10 bits per Hertz.

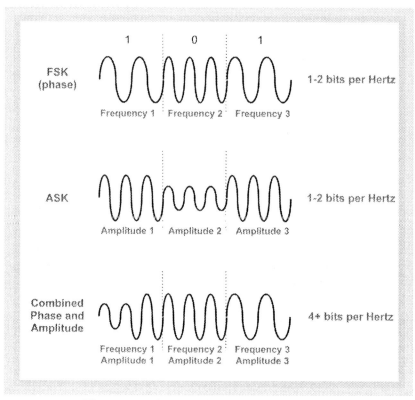

Figure 1.23., Increased Modulation Efficiency

46

Rapid Signal Acquisition

Rapid signal acquisition is the ability of the one device to acquire the signal and/or information of another device. Rapid acquisition is necessary for RF Identification (RFID) tags, wireless automobile tollbooth passes, and interactive devices such as keyboards and mice. The acquisition time for Bluetooth devices version 1.1 can be 10 seconds or more. This is due to the sleep-cycle, signal acquisition time, synchronization time, and information transfer time. Bluetooth acquisition time is likely to be reduced for some or all product types through the use of pilot tones and new signal synchronization procedures.

Index

Printed in the United States
18991LVS00001B/185-202

9 780974 694351

Communication Books
by ALTHOS Publishing

Introduction to Mobile Telephone Systems

ISBN: 0-9746943-2-0   Price: $10.99
Author: Lawrence Harte, Dave, Bowler
#Pages: 48   Copyright Year: 2004

This book explains the different types of mobile telephone technologies and systems from 1st generation analog to 3rd generation digital broadband. It describes the basics of how they operate, the different types of wireless voice, data and information services, key commercial systems, and typical revenues/costs of these services.

Internet Telephone Basics

ISBN: 0-9728053-0-3   Price $29.99
Author: Lawrence Harte   #Pages 226   Copyright Year 2003

Internet Telephone Basics explains how and why people and companies are changing to Internet Telephone Service. Learn how much money can be saved using Internet telephone service and how you can to use standard telephones and dial the same way. Internet telephone service usually costs 1.5 to 4 cents/min for long distance calls and 3 to 10 cents/min for International calls. It describes how to activate Internet telephone service instantly and how to display your call details on the web.

Wireless Dictionary

ISBN: 0-9746943-1-2   Price: $39.99
Author: Althos   #Pages: 628   Copyright Year: 2004

Wireless Dictionary, the Leading Wireless Telephony and Datacom Resource provides over 10,000 of the latest Wireless industry terms and more than 400 illustrations to define and explain latest wireless technologies and services. It provides the references needed to communicate with others in the wireless communication industry where many new terms and concepts are being added each day. It includes directories of magazines, associations and other essential trade resources to help industry professionals find the information they need to succeed in this rapidly growing industry.

Wireless Systems

ISBN: 0-9728053-4-6   Price: $34.99
Authors: Lawrence Harte, Dave Bowler, Avi Ofrane, Ben Levitan
#Pages 368   Copyright Year: 2004

This book describes the different types of wireless communication systems, the basics of how these systems operate, and what services they can efficiently offer. Covered are key market segments for each type of wireless systems along with a description of the leading commercial systems and the services they offer. This book also identifies the key ways these wireless systems are changing to compete with new types of competitors.

Althos Publishing, 404 Wake Chapel Road, Fuquay NC 27526 USA
1-919-557-2260 1-800-227-9681 Fax 1-919-557-2261

Introduction to SIP IP Telephony Systems

ISBN: 0-9728053-8-9   Price: $14.99
Author: Lawrence Harte, Dave Bowler   #Pages: 84   Copyright Year: 2004

This book explains why people and companies are using SIP equipment and software to efficiently upgrade existing telephone systems, develop their own advanced communications services, and to more easily integrate telephone network with company information systems.

Introduction to Signaling System 7 (SS7) and IP, 2nd Edition

ISBN: 0-9746943-0-4   Price: $12.99
Author: Lawrence Harte, Dave Bowler   #Pages: 48   Copyright Year: 2004

Introduction to Signaling System 7 (SS7) and IP control system that is used in public switched telephone networks (PSTN) can be interconnected to other types of systems and networks using Internet Protocol (IP). Some of the interconnection issues relate to how the control of devices can be performed using dis-similar systems (e.g. mixing voice and data systems)

Introduction to Private Telephone Systems

ISBN: 0-9742787-2-6   Price: $11.99
Author: Lawrence Harte,   #Pages: 35   Copyright Year: 2004

Private telephone Systems are communication systems that are owned, leased or operated by the companies that use these systems. They primarily allow the interconnection of multiple telephones within the private network with each other and provide for the sharing of telephone lines from a public telephone network.

Introduction to Public Switched Telephone Networks (PSTN)

ISBN: 0-9742787-6-9   Price: $11.99
Author: Lawrence Harte,   #Pages: 48   Copyright Year: 2004

Public telephone networks are unrestricted dialing telephone networks that are available for public use to interconnect communications devices. There are also descriptions of many related topics, including: Local loops, switching systems, numbering plans, market growth, public telephone system interconnections.

Introduction to Telecom Billing

ISBN: 0-9742787-4-2   Price: $11.99
Author: Avi Ofrane, Lawrence Harte,   #Pages: 36   Copyright Year: 2004

Billing and customer care systems convert the bits and bytes of digital information within a network into the money that will be received by the service provider. To accomplish this, these systems provide account activation and tracking, service feature selection, selection of billing rates for specific calls, invoice creation, payment entry and management with the customer.

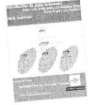

Introduction to Data Networks

ISBN: 0-9742787-3-4   Price: $11.99
Author: Lawrence Harte,   #Pages: 40   Copyright Year: 2004

Data networks are telecommunications networks installed and operated for information exchange between data communication devices such as computers. Because of the dramatic changes that have occurred to Data Networks, such as high-speed data and evolving Ethernet technology.

Althos Publishing, 404 Wake Chapel Road, Fuquay NC 27526 USA
1-919-557-2260 1-800-227-9681 Fax 1-919-557-2261

Telecom Basics, 3rd Edition

ISBN: 0-9728053-5-4   **Price:** $29.99
Author: Lawrence Harte   **#Pages:** 356   **Copyright Year:** 2004

This book provides the fundamentals of signal processing, signaling control, and call processing technologies that are used in telecommunication systems. Covered are analog and digital signal processing, signal modulation, and channel coding. Copper wire, optical transmission systems.

Signaling System 7 (SS7), 3rd Edition

ISBN: 0-9728053-7-0   **Price:** $34.99 **Authors:** Lawrence Harte, Dave Bowler, Richard Dreher,
Toni Beneniger   **#Pages:** 276   **Copyright Year:** 2004

This introductory book explains the operation of the Signaling System 7, (SS7) and how it controls and interacts with public telephone networks and VoIP systems. SS7 is the standard communication system that is used to control public telephone networks.

Patent or Perish

ISBN: 0-9728053-3-8   **Price:** $39.99
Author: Eric Stasik   **#Pages:** 220   **Copyright Year:** 2003

Patent or Perish Explains in clear and simple terms the vital role patents play in enabling high technology firms to gain and maintain a competitive edge in the knowledge economy. Patent or Perish is a Guide for Gaining.

Introduction to Transmission

ISBN: 0-9742787-0-X   **Price:** $11.99
Author: Lawrence Harte,   **#Pages:** 48   **Copyright Year:** 2004

This introductory book explains about the different types of transmission lines, the problems associated with transmission lines, descriptions of solutions to solve transmission problems, and how to control the transmission of information through the use of signaling messages.

Introduction to Wireless Systems

ISBN: 0-9742787-9-3   **Price:** $10.99
Author: Lawrence Harte,   **#Pages:** 48   **Copyright Year:** 2004

This book explains the different types of wireless technologies and systems, the basics of how they operate, the different types of wireless voice, data and broadcast services, key commercial systems, and typical revenues/costs of these services.

Introduction to Switching

ISBN: xx   **Price:** $14.99
Author: Lawrence Harte,   **#Pages:** 48   **Copyright Year:** 2004

Switching systems provide interconnected paths between points or nodes in communication networks. Some of these connections provide for continuous (circuit oriented) communication while others provide for independent bursts (packet oriented) data transmission.

Althos Publishing, 404 Wake Chapel Road, Fuquay NC 27526 USA
1-919-557-2260 1-800-227-9681 Fax 1-919-557-2261

Order Form

Phone: 919-557-2260
800-227-9681
Fax: 919-557-2261
404 Wake Chapel Rd., Fuquay-Varina, NC 27526 USA
email: Success@Althos.com web: WWW.ALTHOS.Com

Date: _____

Name: _____

Title: _____

Company _____

Shipping Address : _____

City _____ State : _____ Zip: _____

Billing Address : _____

City _____ State : _____ Zip _____

Telephone : _____ Fax : _____

Email: _____

Purchase Order # _____ (New account: please call for approval)

Payment (select): Check ___ VISA ___ AMEX ___ MC

Credit Card #: _____ Expiration Date: _____

Exact Name on Card: _____

Qty	BOOK#	ISBN #	TITLE	PRICE EA	TOTAL
Book Total:					
Discounts:					
Sales Tax	North Carolina Residents please add 7% sales tax				
Shipping:	Please apply actual shipping costs and surcharge per above card order				
Total order:					

Printed in the United States
18991LVS00001B/185-202

9 780974 694351